Holt Social Studies

Eastern World
End-of-Year Test
with Answer Key

HOLT, RINEHART AND WINSTON

A Harcourt Education Company

Orlando • **Austin** • New York • San Diego • London

Eastern World   End-of-Year Test

MULTIPLE CHOICE For each of the following, write the letter of the best choice in the space provided.

_______ **1.** The two main branches of geography are
   a. cartography and meteorology.
   b. regional and local.
   c. the study of water and the study of landforms.
   d. physical geography and human geography.

_______ **2.** The main processes of the water cycle are
   a. evaporation and precipitation.
   b. drought and flooding.
   c. salt water and freshwater.
   d. groundwater and surface water.

_______ **3.** Which type of climate can occur at many different latitudes?
   a. tundra
   b. highland
   c. tropical savanna
   d. subarctic

_______ **4.** What is the process of moving from one place to another?
   a. globalization
   b. cultural diffusion
   c. migration
   d. innovation

_______ **5.** Which two physical features gave Mesopotamia its name?
   a. the Fertile Crescent and the Mediterranean Sea
   b. the Persian Gulf and he Mediterranean Sea
   c. the Balkan Peninsula and the Black Sea
   d. the Tigris and Euphrates rivers

_______ **6.** The Resurrection is the Christian belief that Jesus
   a. was crucified.
   b. taught about salvation.
   c. performed miracles.
   d. rose from the dead.

_______ **7.** Which country is made up of a group of islands?
   a. Bahrain
   b. the United Arab Emirates
   c. Qatar
   d. Kuwait

_______ **8.** Which word best describes Central Asia's physical geography?
   a. coastal
   b. fertile
   c. landlocked
   d. tropical

_______ **9.** Which of the following describes why Egypt was called the gift of the Nile?
  a. Egypt was a country of many great rulers.
  b. The Nile brought a humid subtropical climate to the region.
  c. The Nile carried traders from Europe to Asia.
  d. The Nile brought fertility and life to the region.

_______ **10.** The great kingdoms of West Africa became powerful by
  a. controlling trade routes across the Sahara.
  b. establishing colonies in northern Africa.
  c. using ships to send goods around the world.
  d. conquering countries along the Mediterranean Sea.

_______ **11.** Since independence, a problem within many former African colonies has been
  a. conflicts with European countries.
  b. attacks by people from other regions.
  c. fighting among ethnic groups.
  d. high taxes and shortages of workers.

_______ **12.** Both Harappa and Mohenjo Daro were located near the
  a. Arabian Sea.
  b. Thar Desert.
  c. Chang Jiang River.
  d. Indus River.

_______ **13.** Which of the following describes the Qin dynasty under Shi Huangdi?
  a. His policies helped unify China.
  b. His policies led to rebellion and civil war.
  c. His weak leadership resulted in the Warring States period.
  d. China lost territory under his rule.

_______ **14.** What causes tsunamis?
  a. monsoons and typhoons
  b. underwater earthquakes and volcanic eruptions
  c. melting glaciers
  d. deforestation and flooding

_______ **15.** Who were the first humans to live in Australia?
  a. Aborigines
  b. Maori
  c. Dutch settlers
  d. British prisoners

PRACTICING SOCIAL STUDIES SKILLS Study the graph below and answer the question that follows.

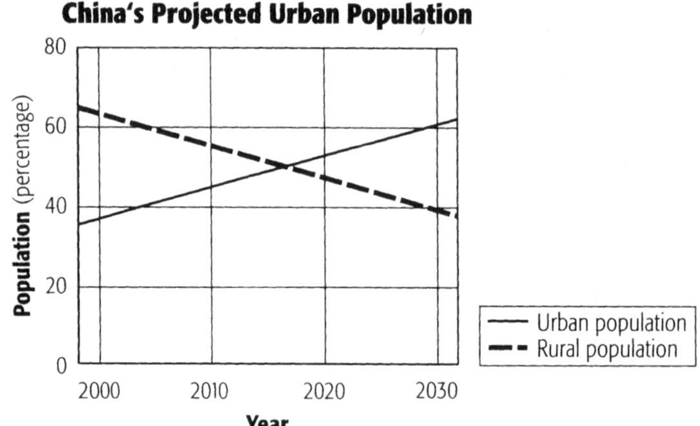

______ **1.** According to the graph, by about how much will the percentage of China's
urban population increase between 2000 and 2030?
   a. 5 percent
   b. 20 percent
   c. 40 percent
   d. 60 percent

FILL IN THE BLANK Read each sentence and fill in the blank with the
word in the word pair that best completes the sentence.

1. Geographers who study the world on a _____________________ level try to
find relationships among people who live far apart. (**local/global**)

2. Earth's _____________________ causes night and day. (**rotation/revolution**)

3. _____________________ is a measure of the number of people living in an
area. (**Population density/Birthrate**)

4. The Bible says that Moses received the _____________________ on a mountain
called Sinai. (**Ten Commandments/Torah**)

5. Desert plains and _____________________ are the main physical features of the
Arabian Peninsula. (**oases/mountains**)

6. Arabs, _______________________, and Soviets all conquered Central Asia and had a major influence on the region. (**Mongols/Samarqands**)

7. After the Kushites were driven out of Egypt, _______________________ became the new Kushite capital and center of a large trade network. (**Meroë/Thebes**)

8. _______________________ ruled Tunisia, Algeria, and parts of Morocco in the early 1900s. (**France/Great Britain**)

9. The _______________________ is a bowl-shaped landform that covers much of Central Africa. (**Congo Basin/Western Rift Valley**)

10. Gupta society _______________________ after the rule of Candra Gupta II. (**began to decline/adopted Hinduism**)

11. _______________________ is green and tropical, with beautiful mountains and crowded cities. (**Mongolia/Taiwan**)

12. Mining is an important industry in the _______________________, Australia's interior. (**Great Barrier Reef/Outback**)

TRUE/FALSE Indicate whether each statement below is true or false by writing **T** or **F** in the space provided.

_______ **1.** Geography is both a science and a social science.

_______ **2.** Earthquakes often occur when two landforms slide past each other.

_______ **3.** A front is a place where two air masses of different temperatures or moisture content meet.

_______ **4.** Babylon was located near present-day Iraq.

_______ **5.** Both Abraham and Moses led the Jews on long journeys.

_______ **6.** The Silk Road was an important route between Europe and Egypt.

_______ **7.** The people of early West African civilizations used written records to preserve their history.

_______ **8.** Europeans built schools and roads in their African colonies and encouraged many Africans to begin and manage their own businesses.

_______ **9.** The lack of major rivers in Central Africa has slowed economic growth.

_______ **10.** The caste system divided Indian society into groups based on a person's birth, wealth, or occupation.

_______ **11.** Korea's main obstacle to reunification is the question of a common form of government.

_______ **12.** The Antarctic Treaty of 1959 designated the continent as an approved site for oil drilling and military activity.

MATCHING In the space provided, write the letter of the term that matches each description. Some answers will not be used.

_______ **1.** fees a country charges on imports and exports

_______ **2.** a regional variety of a language

_______ **3.** placing a limit on trade with another country

_______ **4.** long periods of lower than normal precipitation

_______ **5.** the worship of only one God

_______ **6.** the distance north or south of the equator

_______ **7.** to break away from the main country

_______ **8.** people who want to spread their religious beliefs

_______ **9.** a practice used to dominate other countries' government, trade, and culture

_______ **10.** a landform at the mouth of a river created by sediment deposits

a. missionaries

b. secede

c. drought

d. axis

e. delta

f. polytheism

g. embargo

h. imperialism

i. latitude

j. dialect

k. tariffs

l. monotheism

Progress Assessment

Eastern World

END-OF-YEAR TEST
Multiple Choice

1. d		**9.** d	
2. a		**10.** a	
3. b		**11.** c	
4. c		**12.** d	
5. d		**13.** a	
6. d		**14.** b	
7. a		**15.** a	
8. c			

Practicing Social Studies Skills
1. b

Fill in the Blank
1. global
2. rotation
3. Population density
4. Ten Commandments
5. oases
6. Mongols
7. Meroë
8. France
9. Congo Basin
10. began to decline
11. Taiwan
12. Outback

True/False

1. T		**7.** F	
2. F		**8.** F	
3. T		**9.** F	
4. T		**10.** T	
5. T		**11.** T	
6. F		**12.** F	

Matching

1. k		**6.** i	
2. j		**7.** b	
3. g		**8.** a	
4. c		**9.** h	
5. l		**10.** e	

ISBN 0-03-093460-5

10 1186 16
4500611280

HOLT, RINEHART AND WINSTON
A Harcourt Education Company
Orlando • **Austin** • New York • San Diego • London

APEX
TP_CF